FUTURE FUELS

TO FIGHT CLIMATE CHANGE

by Rachel Kehoe

FOCUS READERS
NAVIGATOR

WWW.FOCUSREADERS.COM

Focus Readers is distributed by North Star Editions:
sales@northstareditions.com | 888-417-0195

Produced for Focus Readers by Red Line Editorial.

Content Consultant: Bingqing Wei, PhD, Professor of Mechanical Engineering, University of Delaware

Photographs ©: Shutterstock Images, cover, 1, 14–15, 17, 19, 22–23, 27; iStockphoto, 4–5; U.S. Department of Energy/Flickr, 7, 21, 29; Horace Hull/Library of Congress, 8–9; Red Line Editorial, 11; Darryl Dyck/The Canadian Press/AP Images, 13; Alexander Shcherbak/TASS/Zuma Press/Newscom, 25

Library of Congress Cataloging-in-Publication Data
Library of Congress Cataloging-in-Publication Data is available on the Library of Congress website.

ISBN
978-1-63739-271-3 (hardcover)
978-1-63739-323-9 (paperback)
978-1-63739-422-9 (ebook pdf)
978-1-63739-375-8 (hosted ebook)

Printed in the United States of America
Mankato, MN
082022

ABOUT THE AUTHOR

Rachel Kehoe is a science writer and children's author. She has published several books and articles on science, technology, and climate change.

TABLE OF CONTENTS

CHAPTER 1

PLANTS INTO ENERGY

A truck arrives at a restaurant in Sacramento, California. It is there to collect used cooking oil. Waste oil is usually thrown out. However, some restaurants give away their used oil. It can be made into **biodiesel**.

The Twin Rivers Unified School District in Sacramento uses biodiesel. The district

Frying oil is a common source of biodiesel.

has adapted 75 school buses to run on this fuel. Biodiesel does more than power the buses' engines. It also helps the planet.

Biodiesel is a type of biofuel. A biofuel is a source of **renewable energy**. Biofuels burn cleaner than gasoline or diesel.

FLEX-FUEL VEHICLES

Some vehicles run on a blend of biofuel and gasoline. One popular type of fuel is called E85. It is 85 percent biofuel. The rest is gasoline. Cars that run on this fuel are called flex-fuel vehicles. A sensor in the fuel tank reads the fuel mix. It sends a signal to another device. That device adjusts how much fuel to send to the engine.

In 2021, more than 4,000 US gas stations offered E85 fuel.

They release fewer harmful gases into the atmosphere.

Another type of biofuel is ethanol. It is made from plants containing sugar. These include corn and sugarcane. Ethanol can fuel regular gasoline-powered cars.

317

CHAPTER 2

THE HISTORY OF ENERGY

Before the **Industrial Revolution**, humans produced little pollution. Wind dried people's clothes. Water from rivers and streams kept people clean. People burned wood for fire. Fire helped cook food and keep people warm. But in the mid-1700s, people started building more and more machines. Railway trains

Early trains burned wood or coal, which made steam. The steam's energy powered the trains' engines.

and automobiles began doing more work. They ran on fossil fuels. These fuels include coal, oil, and natural gas.

Fossil fuels are found all over. They formed from plants and animals that died millions of years ago. Slowly, they became covered in mud and sand. As they broke

RENEWABLE ENERGY

Fossil fuels are a non-renewable resource. Once they are used up, they cannot be replaced. Coal and natural gas are running out. And in 2020, scientists predicted that oil supplies would be gone by the 2060s. But biofuels are a renewable resource. They do not take millions of years to make. Scientists can make biofuels in months.

down, the carbon in their bodies became locked inside. Then Earth's pressure and heat turned them into fossil fuels.

People have burned fossil fuels for hundreds of years. But today, people are using them faster than ever before.

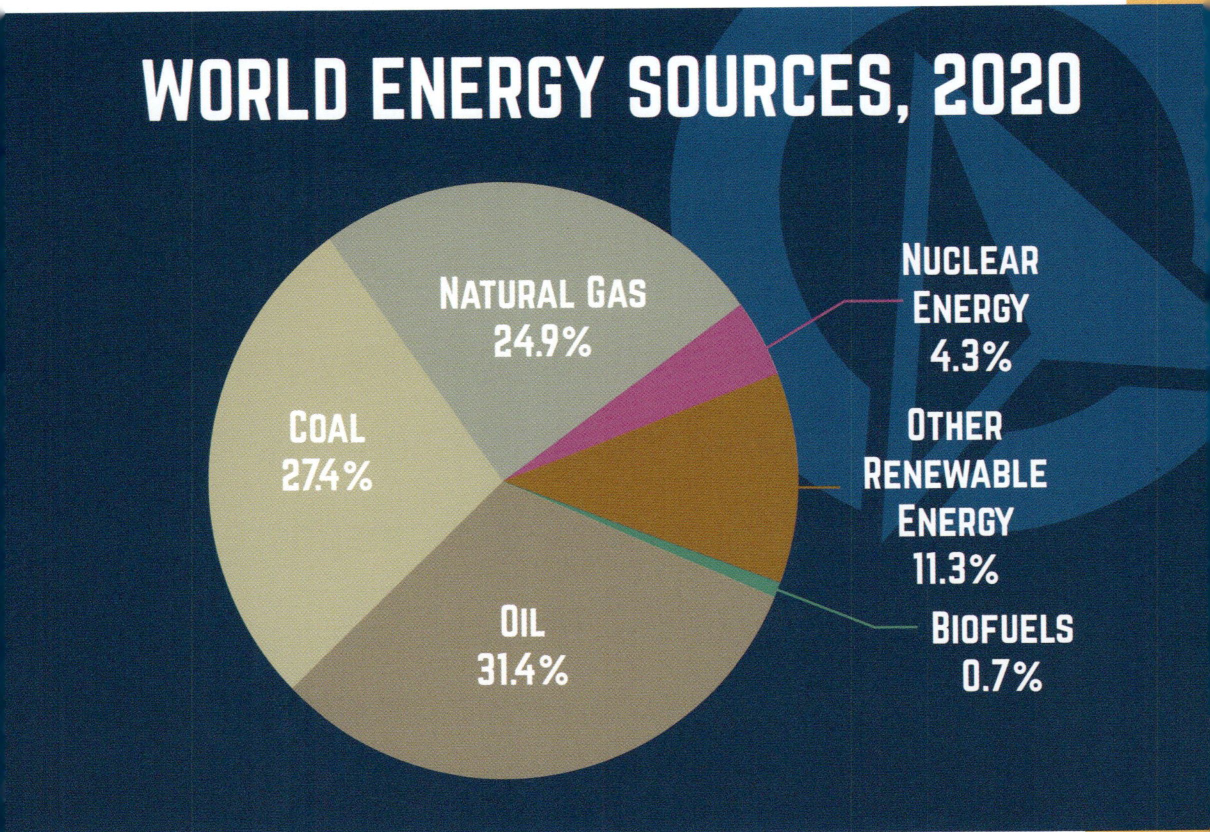

Oil powers cars, trains, and planes. Natural gas heats and cools homes. Coal is burned to make electricity, which powers lights. Electricity also powers TVs, refrigerators, and computers.

Fossil fuels release a burst of energy when burned. But they also release carbon. That carbon bonds with oxygen in the air to make carbon dioxide (CO_2). CO_2 is a greenhouse gas. Greenhouse gases are causing **climate change**. They trap the sun's heat in the atmosphere. Each year, more CO_2 enters the atmosphere. This makes Earth hotter.

Warmer temperatures are causing serious changes around the world. Sea

Lytton, British Columbia, faced record heat in 2021. That heat helped a wildfire burn down the town.

levels are rising. Coral reefs are dying. Glaciers are melting. Extreme weather is also increasing. Floods, droughts, heat waves, and hurricanes all happen more often. These events are getting worse, too. People need a cleaner source of energy. And they need it now.

CHAPTER 3

MAKING BIOFUELS

Biofuels can replace fossil fuels. Biofuels come from the matter of living things. That matter is known as biomass. Examples of biomass include wood, plants, **algae**, and animal manure.

All biomass starts with sunlight. Plants turn the sun's energy into sugar. This process is called photosynthesis. Plants

People have burned wood as a source of biomass energy for hundreds of thousands of years.

use sugar to grow. But the sugar can also make biofuel. Two common types are biodiesel and ethanol.

Biodiesel is a dark-brown liquid. It comes from plant oils. Plant oil is mixed with alcohol. Then it is heated and filtered. Pure biodiesel turns solid in cold

A BIOFUEL HIGHWAY

In 2014, the US government finished a new highway. It was called the I-75 Green Corridor Project. Drivers can travel the highway using only biofuel. The highway crosses six states from Michigan to Florida. Biofuel stations line the highway. Some stations contain ethanol. Others offer biodiesel. The project aims to save 1.7 million gallons (6.4 million L) of fossil fuels each year.

An ethanol manufacturing plant processes sugarcane in Brazil.

weather. As a result, it cannot be used in engines. But scientists solved this problem. They mix biodiesel with fossil fuels. One common blend is B20. This fuel is 20 percent biodiesel. It stays liquid in cold weather. That way, engines run easily.

Ethanol is a clear liquid. It comes from corn and sugarcane. First, the plants go through a process called **fermentation**. This process turns the plants' sugar into alcohol. Next, the alcohol is dried and mixed with gasoline. Ethanol is cheaper than regular gasoline. It also causes less air pollution.

Methane gas from manure is another biofuel. Animals eat plants to get energy. But not every part of the plant is digested. Some pieces end up in the animals' manure. Farmers collect this manure. They feed it into a machine. This machine collects methane gas. It makes a fuel that powers trucks and cars.

Unlike fossil fuels, biofuels are **carbon neutral**. When plants grow, they take in carbon from the air. When burned, the fuels release that carbon back. Biofuels give off only what they have taken. No new CO_2 is released. As a result, using biofuels can help slow the climate crisis.

CARBON-NEUTRAL BIOFUELS

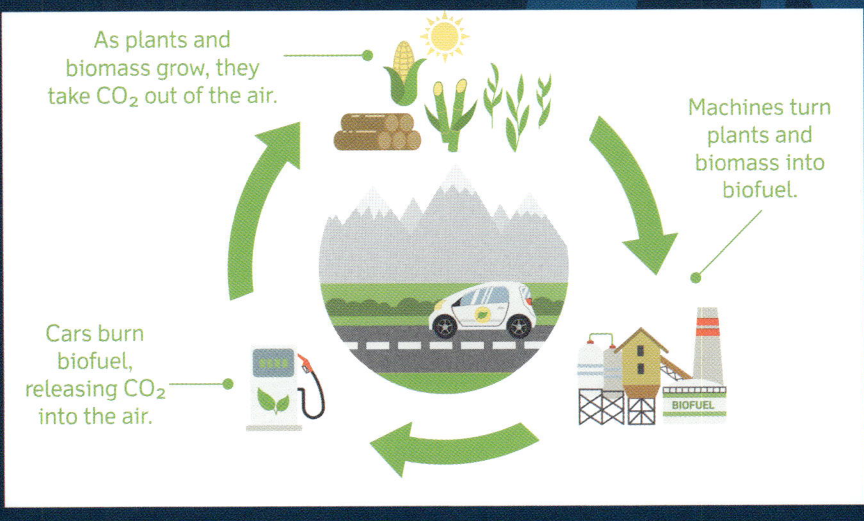

HYDROGEN ENERGY

Hydrogen is one possible energy source for the future. This gas can fuel cars and other vehicles. These vehicles often use fuel cells. These cells use hydrogen to make electricity.

First, a driver goes to a hydrogen fueling station. He or she fills the car's fuel tank with hydrogen gas. When the car runs, gas is sent from this tank to fuel cells. In each cell, hydrogen combines with oxygen. This makes electricity. The electricity powers the engine. In this way, fuel cells are similar to batteries. But unlike batteries, they don't die. They also don't need to be recharged. Fuel cells just need to be refueled.

Using hydrogen doesn't create pollution, either. It releases only heat and water. For this reason, the gas could help replace fossil fuels.

One way to make hydrogen uses sunlight to split apart water. The hydrogen separates and forms bubbles.

However, hydrogen energy faces other challenges. In the early 2020s, most hydrogen came from natural gas. As a result, making hydrogen still released lots of greenhouse gases. Even so, scientists were working on cleaner ways of making hydrogen.

CHAPTER 4

FUELING CHANGE

Like all energy sources, biofuels have problems. Burning biomass creates less pollution than using fossil fuels. But it is not as clean as wind and solar energy. Scientists see biofuels as a temporary solution. These fuels can give people power until cleaner sources are ready.

All cars that burn fuel produce some amount of pollution.

Electric cars already exist. They release no CO_2 as they run. But as of the early 2020s, most cars were not electric. In the meantime, biofuels can meet many people's energy needs. People do not need to depend on only fossil fuels for power.

Some cars and trucks already run on biofuels. However, far fewer planes and ships have switched away from fossil fuels. Biofuels for airplanes and ships are expensive. Scientists continue working to make cheaper options.

Biofuels have other disadvantages. Growing and harvesting crops takes energy. Most of that energy comes from

Cleaner fuels for planes are known as sustainable aviation fuels.

fossil fuels. Crops also use lots of water. They need fertilizers and pesticides as well. These chemicals can pollute waterways. Some fields are burned after harvesting. This causes air pollution.

Growing crops for fuel also takes up land. Critics argue that this land could be used to grow food. Other farmers may turn forests into fields. Cutting down native forests hurts plants and animals. Plus, removing trees means less CO_2 is taken out of the air. This can make climate change worse.

For these reasons, scientists are looking at new ways of making biofuels. One example is **cellulosic ethanol**. This fuel provides more energy per plant. No extra water or land is needed. Cellulose is a sugar that comes from the woody parts of plants. Switchgrass and wood chips are popular sources. However, removing this sugar is difficult. Scientists are looking for easier ways to break it down.

Cellulose can come from stems, stalks, leaves, and husks.

Algae provide another good source of energy. These organisms are found in waterways. Algae make sugar using the sun's energy. Then they convert it into oil. The oil from algae can make biodiesel.

Algae are easy to grow. They make more oil than many plants do. This ability makes algae a good option for biofuel. Algae also thrive in harsh conditions. They grow where most plants cannot. This means they do not compete with food crops. Scientists are working to

BRAZILIAN BIOFUEL

In 1973, gasoline prices shot up. In response, Brazil decided to grow its own fuel. Farmers planted more sugarcane. This crop contains lots of sucrose. This type of sugar easily turns into ethanol. By 1985, ethanol powered nearly every vehicle in Brazil. But in 1986, gasoline prices dropped. Sugarcane crops failed. As a result, ethanol became costly. So, Brazil introduced flex-fuel cars in 2004. As of 2021, most Brazilian cars used a mix of fuels. This shift reduced the country's need for fossil fuels.

A scientist studies how to grow algae for biofuels.

make more algae. They want the algae to have as much oil as possible. However, removing the oil is challenging. Research into algae fuel continues.

More people are making the switch to biofuels. Their choices will help meet the energy needs of the future.

FOCUS ON
FUTURE FUELS

Write your answers on a separate piece of paper.

1. Write a sentence that describes the main ideas of Chapter 3.

2. Would you want to ride in a vehicle that uses biofuel? Why or why not?

3. Flex-fuel vehicles use a mix of ethanol and what other fuel?

   A. gasoline
   B. biodiesel
   C. hydrogen power

4. Why do scientists believe biofuels are a temporary solution?

   A. Biofuels take millions of years to replace.
   B. Biofuel vehicles pollute more than electric vehicles.
   C. Biofuels are not carbon neutral.

Answer key on page 32.

GLOSSARY

algae
Tiny plant-like organisms that produce oxygen.

biodiesel
A renewable fuel made from plant oils.

carbon neutral
When something releases the same amount of carbon dioxide as it absorbs.

cellulosic ethanol
Ethanol made from the tough parts of green plants, including stems and stalks.

climate change
A human-caused global crisis involving long-term changes in Earth's temperature and weather patterns.

fermentation
A process in which proteins break down plants and convert their sugar into alcohol.

Industrial Revolution
Starting in Great Britain in the 1700s, a huge economic shift involving the use of powerful machines and mass production.

renewable energy
Energy produced from a source that will not run out.

TO LEARN MORE

BOOKS

Eboch, M. M. *The Future of Energy: From Solar Cells to Flying Wind Farms*. North Mankato, MN: Capstone Press, 2020.

Hirsch, Rebecca E. *Climate Change and Energy Technology*. Minneapolis: Lerner Publications, 2019.

MacCarald, Clara. *Energy from Algae*. Lake Elmo, MN: Focus Readers, 2022.

NOTE TO EDUCATORS

Visit **www.focusreaders.com** to find lesson plans, activities, links, and other resources related to this title.

INDEX

Answer Key: 1. Answers will vary; **2.** Answers will vary; **3.** A; **4.** B